QUALITY ASSURANCE FOR
THE CHEMICAL AND PROCESS INDUSTRIES
A Manual of Good Practices

Prepared by

American Society for Quality Control
Chemical and Process Industries Division
Chemical Interest Committee

AMERICAN SOCIETY FOR QUALITY CONTROL
310 West Wisconsin Avenue
Milwaukee, Wisconsin 53203

QUALITY ASSURANCE FOR THE CHEMICAL AND PROCESS INDUSTRIES
A Manual of Good Practices

Prepared by

American Society for Quality Control
Chemical and Process Industries Division
Chemical Interest Committee

AMERICAN SOCIETY FOR QUALITY CONTROL
310 West Wisconsin Avenue
Milwaukee, Wisconsin 53203

ISBN 0-87389-035-3

Table of Contents

Preface

The Chemical and Process Industries Division of the American Society for Quality Control has prepared this manual in view of the recent revival of interest in quality for both the supplier and the customer. A technical subcommittee was formed to address the various issues and to provide the practical experience from 12 corporations. The writing committee consisted of David Anderson, Gary Anderson, Jim Bigelow, Bradford Brown, Georgia Kay Carter, Ken Chatto, Louis Coscia, David Files, Peter Fortini, Richard Hoff, Sherman Hubbard, Rudy Kittlitz (chairman), Norman Knowlden, Anil Parikh, David Stump, and Jack Weiler. The corporations were: Allied-Signal Inc., American Cyanamid Company, Dow Chemical Company, Dow Corning Corporation, E.I. du Pont de Nemours & Company (Inc.), Eastman Kodak Company, Exxon Chemical Company, Hercules, Inc., Omni Tech International Ltd., Shell Chemical Company, 3M Company, and Uniroyal Chemical Company. Rad Smith of the Ford Motor Company added constructive comments and Faye Carey of E.I. du Pont de Nemours & Company (Inc.) gave administrative support.

1. Purpose

This manual documents good quality practices in the chemical and process industries (CPI). It is intended for quality professionals, process engineers, manufacturing managers, suppliers, and customers of the chemical and process industries to use as a reference to build, guide, or evaluate the suitability of a quality system.

Perspective:

The art and science of sampling and measurement are different in chemical and process industries compared to mechanical industries. The relatively large contribution of measurement error requires proper use of statistical methods for process control. Because of this, proper interchange and understanding of information between customers and suppliers becomes very important.

Additional considerations requiring special attention and a different framework to apply statistical methods in the chemical and process industries are listed as follows:

- Raw materials are often natural materials whose consistency depends on natural forces. This requires making compensating adjustment in the process to maintain a consistent quality level of the finished product.
- The technology of chemical sampling presents complex problems because of chemical considerations and the varied nature of product units ranging from a few grams to thousands of tons.
- Chemical processes occur on the molecular level and changes are often inferred from secondary information.
- Measurement itself is a complex process which must be carefully standardized and controlled.

1

- Automatic process control is highly developed with computerized feedback and feed-forward loops. Emphasis is on "prevention."

- Once produced, product properties often change with time as well as from changes in environmental conditions.

- There may not be a direct relationship between the measured properties and the performance characteristics of the products in use.

- True defectives are rarely found, although nonuniformity between lots is often detected by the customer.

- Health, safety, and environmental impact of both the process and the product are major concerns.

We have tried to focus on these issues and to balance them with the needs of the industry and its customers for sound quality practices. This is not a "stand alone" document and a bibliography has been included.

2. Organization and Responsibilities

Scope:

- A company should document its quality policy, quality system, and quality planning requirements in a manual or series of manuals. The Quality Manual should clearly state policy, define procedure, and specify responsibility for quality system components.

- A company should make internal quality reports periodically to show progress toward meeting quality objectives.

- Constant innovation and superior customer service create excellence. Organizations must learn to develop flexibility while maintaining consistency. Innovation and quality improvement are different processes from quality control and result in change rather than maintenance of status quo.

- Health, safety, and environmental impact of both the process and the product are major concerns in the chemical industry. A company should document its safety policy and procedures in a manual or a series of manuals which specifies responsibility for safety system components.

2.1 Quality Policy

The following has been taken from "Quality Management and Quality System Elements — Guidelines," ANSI/ASQC Q94-1987, Sections 4.1 and 4.2:

> "The responsibility for and commitment to a quality policy belongs to the highest level of management. Quality management is that aspect of the overall management function which determines and implements quality policy. The management of a company should develop and state its corporate quality policy. This policy should be consistent with company policies. Management should take all necessary measures to ensure that its corporate quality policy is understood, implemented, and maintained."

2

2.2   Quality Systems

The sharing of product data with the customer should not take place unless there is a functioning management of quality systems in place.

Management of quality systems involves:

- Top-down management participation.
- Statistical process control.
- Knowledge of measurement variability, process variability, and "tools" to keep them under control.
- Documented procedures for consistent treatment of information, i.e., data.
- A program for continual improvement.

2.3   Quality Manual

The quality manual should include the following:

2.3.1   A description of how quality will be designed into the manufacturing process and product, including:

- Raw material control systems.
  - — Vendor selection and certification method
  - — Lot sampling and acceptance practices
  - — Test methods
- Process control systems and practices.
- Testing, analysis, and product acceptance systems.
  - — Good laboratory practices
  - — Documented test methods
  - — Sampling plans
  - — Sampling procedures
  - — System for maintaining standards
  - — Standardization/calibration
  - — Lot acceptance and product control systems
- Product shipment systems.
  - — Packaging and labeling quality assurance
  - — Shelf life control
  - — Shipment container quality assurance
- Nonconforming material disposition.
  - — Recall/retrieval practices
  - — Nonconforming raw material
  - — Nonconforming product

3

2.3.2 A definition of customer relations practices.

- Policy for sharing information.
- Method for analyzing and resolving complaints, including corrective action procedures.
- Method for determining customers' needs.

2.3.3 A description of regulatory matters that may impact company operations.

2.3.4 Documentation of procedures for developing and commercializing new products.

2.3.5 A description of the standard operating procedure and its administration.

2.3.6 A definition of record retention practices.

2.3.7 Organization charts of the company and of the quality function for assistance in contacting company personnel.

2.3.8 Specifications.
- Raw materials
- Process
- Product
- Package and label
- Sales

2.3.9 A description of the quality audit system and practices.

2.4 Quality Reporting

2.4.1 Process quality data for the purpose of control should be available to operating personnel, preferably on a real-time basis. Summary reports of process performance should be available for line management to provide awareness, identify improvement opportunities, and allocate resources.

2.4.2 Reports should be made to management showing progress against company quality objectives. Areas for reporting include customer perception of quality and value, controlling process to target improvements, and process capability improvements. The purpose of management reports is to inform and to assist in the allocation of resources. Quantitative measures should be used whenever possible.

4

2.5 Quality Improvement

2.5.1 Quality improvement.

- Assessment
  - Existing level and variability of process (used in its broadest sense) should be measured statistically over time.
- Process control and correction
  - Remove error and inconsistency (special causes).
  - Reduce variability. Emphasize on-line measurement and on-line control.
  - Maintain closeness to target.
  - Change process level through process improvement.
- Process improvement
  - Quality improvement must be led by management and must involve all employees.
  - Quality improvement requires formal periodic review, assessment, and organization to develop that improvement.
  - Quality improvement occurs only when process knowledge is increased. Procedures such as experiment design, screening studies, evolutionary operations, analysis of variance, analysis of means, etc., should be used to increase process understanding.
  - Quality improvement occurs when teams, trained in problem solving and statistical tools, cut across departmental barriers and work together.

2.5.2 Customer service and product performance.

- Superior customer service requires a marketing system which can determine and transfer customer needs and desires into actual product performance and customer service.
- Process data collection and analysis should ultimately be integrated into systems linking it to the product's field performance.

2.5.3 Innovation.

Innovation requires a stimulating atmosphere and an environment that can transfer innovation into the organization. Innovation requires flexible systems, cross-functional teams, technical knowledge, and information-handling skills.

2.6 Safety

The safety manual should include a company's overall approach to safety management:

- Operational guidelines to handle all processes and products in a safe manner.
- Safety rules for all manufacturing and nonmanufacturing areas.
- A description of safety audit system and practices.
- A description of regulatory matters that may impact company operations.
- An approach to provide pertinent safety information to company product users.

3. Basis for Specifications

Specifications are the agreed upon, documented requirements between customer and supplier. They should include sampling and testing methods, limits, targets, and reporting requirements.

Scope:

- Guidelines for determining product specifications should be provided. Whenever possible, specifications should be in terms of true value, i.e., the required property value if measured without measurement variability.
- Good quality practice should be documented concerning retention of records related to the quality of a product.
- The purpose of obtaining a sample from a production lot is to provide information for an effective decision regarding acceptance or rejection using the lot acceptance plan. The sample should be representative of the entire lot. It can be either a random sample or a systematic sample. In general, it is preferable to have a minimum sample size of two, with the samples taken at distinctly different locations in the lot. This will allow estimates of the within-lot variability to be used in designing lot acceptance plans. Caution must be used to ensure that the samples are independent and represent within-lot variance and not just measurement error.
- Good quality practice should be documented concerning interpretation of numerical data and specification tolerances, which are the essential elements for agreement on the critical measurements for product acceptance. These are based on the general procedure and conventions of Food Chemical Codex, United States Pharmacopeia, National Formulary, American Society for Testing and Materials, and American Chemical Society (Reagent Chemicals).

3.1 Definition of Terms

3.1.1 *Specification.* A precise statement, or set of requirements (quantities, characteristics, or properties) to be satisfied by a material, process, or product with notations as appropriate of the procedure(s) which should be used to determine if the requirements are satisfied. As far as practicable, each requirement should be expressed numerically as a target value with required limits.

3.1.2 *Target.* Agreed upon value within the specification limits that produces an optimal result for the customer and/or supplier. The target need not be the center of a specification range.

3.1.3 *True value.* A quantitative characteristic that does not contain any sampling or measurement variability.

3.1.4 *Product variability.* Amount of variability in a set of measured values due to actual differences in product characteristics from product unit to product unit or from time to time. From the customer's viewpoint, the supplier's product variability manifests itself as variability in the performance of the customer's process performance as the customer consumes the material.

3.1.5 *Purchase specification.* A formal document stating all of the information necessary for the purchase of a process material, including:

- Specifications (target value and limits) for all properties which define the required product quality. The specific test method should be specified.
- Packaging and shipping requirements.
- Safety, health, and environmental requirements.
- Applicable federal and state regulations.
- Terms for rejection of material.

3.1.6 *Sales specification.* A formal document containing the information necessary to describe a product for a particular customer-use and end-use.

3.1.7 *Lot.* A definite quantity of a product or material accumulated under conditions considered uniform for sampling purposes. For a continuous process, a lot is usually defined by a time interval. For a batch process each batch can be considered a lot. This definition is based on the broad definition of a process as being any activity which changes the quality parameters used to define the product: blending, mixing, packaging under different conditions, etc., as well as the usual production processes.

3.1.8 *Process capability.* The variability of a product property when the process is in a state of statistical process control.

7

3.1.9 *Process performance.* The variability of a product property as produced over an extended period of time. Process performance consists of process capability plus all other sources of variability experienced during the time period.

3.2 Determining Product Specifications

3.2.1 Product specifications.

- Whenever possible, specifications should be stated in terms of true product variability; i.e., the required property values if measured without measurement variability.

- Customer requirements, if known or determined by technically sound procedure, should be the primary basis for specifications.

- If customer requirements have not been determined, or if customer and supplier targets differ, specifications must be developed and agreed to by the supplier and customer.

- The supplier's demonstrated process performance should be the starting point for specifications. The customer's experience with this product should also be considered.

- If the customer's experience has been satisfactory, specifications should be set close to the supplier's demonstrated performance.

- If the supplier's process has not been producing satisfactory product, specifications should be agreed to that the supplier can meet and the customer is willing to accept. The supplier and customer should also agree on goals and timing for supplying satisfactory product. This may involve process improvements or the use of a lot acceptance plan for product release.

- Temporary specifications should be agreed to for new products or when there is insufficient supplier or customer data available to determine specifications. When sufficient data is available, specifications should be reviewed for any required changes.

3.2.2 Changing specifications.

- Specifications based entirely on customer requirements should be changed only when customer requirements change.

- Negotiated specifications should be changed as follows:
  - Specifications based on process performance may be reviewed for change when process performance improves or when customer requirements have been determined.
  - Specifications determined from insufficient data should be reviewed for change when sufficient data becomes available.

- Specifications should be reviewed for change when suppliers or customers make any change in their processes that could affect the adequacy of current specifications.

3.3 Incoming Material Acceptance and Product Release

3.3.1 Incoming material acceptance.

Incoming material acceptance procedures should be developed, taking into consideration process requirements and supplier capability. The goal is to minimize or eliminate the need for formal incoming lot acceptance and to rely on the supplier to supply products that continually meet specifications. Supporting documentation may be required in the form of certification, data sharing, or statistical evidence. Minimal inspection or testing may still be required for identification or to detect changes occurring during shipment.

Formal lot acceptance procedures may be required based on prior experience with the material or because there is a need for risk avoidance.

3.3.2 Lot acceptance for product release.

If the existing process control is not sufficient to ensure shipping products that continually meet specifications, it may be necessary to use a lot acceptance procecdure.

3.4 Sampling

Samples taken from incoming materials acceptance may be for identification only or for lot acceptance against specifications. Samples taken from in-process materials or from final product may be for process control or for lot acceptance against specifications or both. When samples are used for both process control and product release the sample frequency is then determined by the most critical application. When the sampling is done for the purpose of lot acceptance, each lot should be sampled by taking a representative sample either randomly or systematically. In general, it is preferable to have a minimum sample size of two, taken from distinctly different locations within the lot; this will allow ongoing estimates of the within-lot variability used in designing effective lot acceptance plans.

3.4.1 Discrete units.

When the lot is made up of discrete units such as bags, drums, bales, or tubes of yarn, the sample should consist of material from at least two different units. A random sample is often adequate for incoming lots of material. However, a systematic sample is preferred when the units are in some form of a time sequence, as in most product being released for shipment. The systematic sample should be distributed throughout the lot so that the within-lot variability is included.

3.4.2 Bulk product.

When the lot is a single unit of bulk material such as high or low viscosity liquid, powder, pellets, or particles, the sample should be taken from at least two locations in the lot. The sample locations can be chosen randomly, and the individual samples may be grab samples, thief composite samples, or stratified samples; i.e., top, middle, and bottom.

3.4.3 Blends.

Blends apply primarily to product lots being released for shipment. Blends are used to combine material from two or more in-process sources for efficient utilization of equipment.

- If blending is used to reduce the effect of in-process differences, good quality practice dictates modifying the process to eliminate these differences so that blending is not required.

- If specifications are on the blended product, the sample should be taken on the blended material.

3.5 Measurement Variability and Acceptance Decisions

3.5.1 Acceptance decisions are made based on observed or measured value. The observed value is the sum of a true value and a random error from the measurement process. The total variance of observed values is equal to the variance of the true values, plus the variance of the random measurement errors. In the CPI, measurement variability is often a large proportion of the total observed variability. It is not uncommon for the measurement variance to be 50 percent of the total observed variance.

3.5.2 Industry accepted test methods are preferred whenever possible. In any case, when selecting the test method to be used, the effect of measurement variability on the reported result should be considered. Measurement variability can be viewed as consisting of repeatability and reproducibility. Repeatability refers to the variability in the reported results measured at the same time, on the same equipment, by the same operator, etc. Reproducibility refers to the additional variability in the reported results measured at different times, on different equipment, by different operators, etc.

3.5.3 Replicate testing can reduce the effect of measurement variability on the final reported result. If one replication is done within a short period of time, such as one after another, only the part of the measurement variability associated with that time period is reduced. However, if a large source of measurement variability is from day-to-day, the effect of this variability can only be reduced by replication over days. A time series component of variance analysis may be useful in planning the replication schedule.

3.5.4 The amount of measurement variability affects decisions made from measurement results. This is particularly important in the CPI.

Limits used to decide if product meets specifications should be determined using a procedure based on the Operating Characteristic curve (OC curve) of the test. The shipping and/or acceptance limits can then be selected, balancing the risks of accepting materials which do not meet specifications and rejecting the materials which meet specifications. Replications can be used to change the shape of the OC curve and thus change the risk.

3.5.5 When the test method includes a precision statement, customers using well-calibrated equipment should accept the product unless their acceptance values exceed the supplier's reported value by more than expected, based on the precision statement.

3.6   Acceptance Plan Design

- Acceptance plans for either incoming materials or product release use the same design criteria. The acceptance decision is made on a lot-by-lot basis. For incoming materials, an entire shipment may be a single lot, or the shipment may be divided into one or more lots for acceptance purposes. For products to be shipped, the lot may be a single unit or a predetermined number of units.
- Acceptance plan design (sample size and limits) must take into account the effect of within-lot and measurement variability. This is of utmost importance when the measurement variability is a large portion of the total observed variability. This will involve making use of the Operating Characteristic (OC) curve for each plan.
- Standard acceptance plans may be found in the following references:
  — ANSI/ASQC Z1.4 - 1981. Sampling Procedures and Tables for Inspection by Attributes.
  — ANSI/ASQC Z1.9 - 1980. Sampling Procedures and Tables for Variables for Percent Nonconforming.

Standard plans may not be readily adapted to many chemical operations.

3.7   Operating Characteristic Curve

The OC curve is a plot for a specified acceptance sampling plan that shows the probability of accepting a sampled lot for any given true level of the parameter being measured. The parameter being measured may be either an attribute or a variable.

The OC curve is used to select acceptance limits and a sample size that allows for within-lot and measurement variability with predetermined risks of accepting material outside of specifications and rejecting material that meets specifications.

3.8   Risks of Error

Alpha risk is the probability of making a Type I error, i.e., rejecting a lot which is "good" or within specifications. This is sometimes referred to as the producer's risk.

Beta risk is the probability of making a Type II error, i.e., accepting a lot which is "bad" or outside of specifications. This is sometimes referred to as the consumer's risk.

3.9   Record Retention

3.9.1   Retention time.

All data related to the lot of any product produced for sale or internal use shall be maintained for whichever of the following is longer:

- The stated useful life of the specific lot.
- One year after the final sale of the specific lot.
- Regulator requirements (e.g., FDA).

All information not related to a specific product should be maintained for a period of two years.

3.10   Number Rounding

3.10.1   Rounding of numerical data.

- To round a number to a given number of significant figures, consider the digit in the next place beyond the one to be retained. Then follow rounding procedures as described in ASTM Standard E 380, Section 4.4, Rounding Values, Section 4.5, Conversion of Linear Dimensions of Interchangeable Parts — Method A, and ASTM Standard E 29, Section 3. In a string of calculations, do not round until the last step.

3.10.2   Significant figures for tolerance limits.

- When limits are expressed numerically, the upper and lower limits of a range are inclusive so that the range consists of the two values themselves and all the intermediate values, but no values outside the limits.

- For specifications, values are considered significant to last digit shown. For example, a requirement of not less than 96.0 percent would be met by a result of 95.96 percent but not by a result of 95.94 percent. Similarly, a specified range of 46 percent to 49 percent means only values between 45.5 and 49.4 will comply (following rounding procedures as in Section 3.10.1).

3.10.3 Examples of rounding off.

Acceptance Limit	Observed Value	If Rounded off to Nearest	Correct Rounded off Value	Complies with Acceptance Limit
60,000 psi, min.	59,940	100 psi	59,900	No
	59,950	100 psi	60,000	Yes
	59,960	100 psi	60,000	Yes
57%, min.	56.4	1%	56	No
	56.5	1%	56	No
	56.6	1%	57	Yes
0.5%, max.	0.54	0.1%	0.5	Yes
	0.55	0.1%	0.6	No
	0.56	0.1%	0.6	No

3.10.4 Unit conversion in measurement systems.

- Accurate conversions are obtained by multiplying the specific quantity by the appropriate conversion factor given in the ASTM Standard E 380 or the National Bureau of Standards (NBS) Handbook 44, 1979.
- In terms of significant figures the rule for multiplication and division is that the product or quotient shall contain no more significant digits than are contained to the right of decimal in the number with the fewest significant digits used in the multiplication or division.
- Example of conversion:

$113.2 \times 1.43 = 161.876$ which rounds to 161.9.

4. Sampling Technology

Scope:

Sampling is the process of removing a portion of a material for analysis. Ideally, the sample should truly represent the material from which it was taken. It must maintain its integrity with respect to the characteristics measured, and the act of sampling should not change the chemical sampled. Sampling should provide low cost measurement data on the material with known and controlled risks.

Sampling technology involves discussion of the sources of sampling variation with respect to making effective decisions.

The physical state of the material affects the sampling process.

The environment, including containers, of the material to be sampled must be considered in order to protect samplers from the chemical and to protect the material from contamination, change, or degradation during sampling.

Samples may be taken for lot acceptance or release, for process control, and for many other reasons. The reason for taking the sample as well as the physical state, the toxicity, the source, the measurement variability, the length of the analysis, and many other parameters must be considered in determining a sampling plan.

The sampling approach may differ, depending on whether the characteristics measured are attributes or variables. Sampling choices include 100 percent inspection, sampling plans, or no sampling.

Training is required to develop personnel with the skills and knowledge needed to take samples.

Retaining samples is generally a good practice, but except for certain regulatory requirements, sample retention is not an absolute requirement. There is a balance between retaining all samples for a long time to cover all eventualities and the real economic problem of storage space, especially when a special environment is required.

4.1   Sampling Variation

4.1.1   Definition of population.

- In the chemical and process industries the population sampled for product acceptance is usually a lot (see Section 3.1.7).
- For process control the population sampled is generally a function of process time.

4.1.2   Estimation-confidence.

- Decisions regarding process control or product acceptance depend upon the homogeneity of a characteristic measured for a lot.
- The level of the characteristic for the lot is estimated as the average of the sampling measurements of that characteristic taken on that lot. The samplings must be sufficient to represent the variability of that characteristic throughout the lot.
- The homogeneity in the lot is estimated from the variability found between measurements on the representative samplings taken and from the sampling variation.
- Increasing the number of samplings taken and measured will provide greater confidence in estimating the level and in estimating the

sampling variance. (An increase in sampling does not necessarily decrease the sampling variance.)

- Increasing the number of measurements taken on each sampling can be used to reduce the contribution of measurement error to the overall sampling variance.

4.1.3 Sampling variation.

- Between samplings

  The variability detected between measurements of a characteristic on replicate samplings of a lot provides a measure of the lot homogeneity, within-sample variability, and measurement variability.

- Within a sampling

  Replicate determinations of a characteristic on the same sampling provide a measure of the within-sample variability and measurement variability.

4.1.4 Homogeneity considerations.

- Process knowledge

  — Knowledge of the chemical(s) and process involved should be used to select the proper sampling technique to minimize sampling costs and error.

  — Supporting information on lot formation, process capability, and run order, as well as supplier certification, process control systems, and quality audit results should also be used.

- Discrete units

  The representative sample should consist of material obtained from discrete units distributed throughout the lot so that within-lot variability is included in the sample.

- Bulk material

  The representative sample should consist of material obtained from locations distributed throughout the lot so that within-lot variability is included in the sample.

  — Within-lot variability tends to increase as one proceeds from single ingredient gases through low viscosity liquids; solutions; single ingredient uniform size solids and on to high viscosity liquids; emulsions and multiple particle size solids, mixes and up to coal and ore. In turn the number of lot locations sampled should increase accordingly.

4.1.5 Composite samples.

When samples are composited and characteristics measured on replicates of the composite sample, information on within-lot variability is lost and only measurement (and sample pooling) error remains.

4.2 Method

There should be a written description of the sampling process and a statement of the use to be made of the results for every sampling operation. The sampling method description should contain:

- A description of appropriate environment required to maintain the quality of the material and its samples.

- A listing of sampling equipment and container requirements.

- A written procedure on how the sample is to be obtained, in order to ensure consistency.

- A definition of the exact location of each sampling point.

- A statement of the sample size needed for testing, both weight or volume and subgroup. Note that multiple samples are not always economically or physically feasible.

- A list of the analyses to be performed.

4.3 Physical State and Source

4.3.1 The physical state describes the state of the material as a liquid, gas, solid, or mixture of these states; and also designated conditions such as high or low viscosity, powder, pellets, flakes, slurry, emulsion, or particles. The source of the material to be sampled can be in pipes, conveyors, piles on the ground, storage bins, tanks, rail cars, tank wagons, drums, bottles, bales, etc. Both the physical state and the source affect the sampling equipment used, the equipment needed to protect the material from the environment, the homogeneity of the material, and the number of samples required.

4.3.2 In a system containing particulates, it is desirable to take samples when the system is in motion.

4.3.3 Sampling systems should be designed to avoid "dead legs"; i.e., fresh, current material must be dispensed by the sample system, not material that has been sitting in the system.

4.4   Environment

4.4.1   The proper environment includes provision for health, safety, and other considerations which impact on the quality of the material in either its bulk or sample form. Examples: Toxic chemicals, chemicals absorbed through skin, chemicals under high pressure, and chemicals sensitive to air, moisture, heat, or cold.

- The sampler must be protected from exposure to the material and also needs to be aware of potential hazards associated with each chemical sampled.
- The sampled material must be protected from exposure to contamination.
- The sampled material must be protected from exposure to an incompatible environment.

4.4.2   Container

- Material of composition

  The container or its closures must neither react with nor be soluble in the sample. Examples: Glass is generally inert, but soft glass can contaminate a sample with sodium; plastics have additives which can affect samples; solvents can permeate polyethylene; caps, covers, and cap liners can also be sources of contamination.

- Cleanliness

  Cleanliness appropriate to the analysis is mandatory. New containers may contain fibers, dust, and dirt. Washed containers may contain detergent residues. Tap water can contaminate certain samples; materials of construction for equipment, and piping or tubing for distilled or deionized water can also be a contamination source. Appropriate conditioning and/or purging of sample lines and containers may be required before taking the sample.

4.4.3   Equipment

Sampling equipment is an ever-important consideration. Construction, design, and suitability of material must be carefully considered.

4.5   Sampling Plan

4.5.1   Lot acceptance or release.

Published sampling plans, such as ANSI/ASQC Z1.9-1980, are for lot acceptance and release and may be suitable for chemical sampling. As process control and process capability approach new levels of performance, the traditional sampling plans become less useful.

4.5.2   Process control.

For process control, especially continuous processes, it may be advisable to take samples at specified time intervals. For batch processes, sampling may be made at certain stages in the process. Process control sample frequencies depend on the process capability, the repeatability of the analysis, the residence time of the process, the time elapsed for the analysis, the cost of the analysis, economics, and the risk factors involved. These generally have to be considered on a case-by-case basis; rarely do standard sampling plans apply.

4.6   Sampler Training

Primary to sampling chemicals is an understanding of the chemistry behind a sampling method and development of the skills needed to use the tools of sampling. A general understanding of the statistical theories, and how, when, and why they apply is also an important part of a sampler's training. The degree to which a person understands the chemistry and statistics, and develops the necessary tool skills should be measured. Appropriate levels of performance and training need to be established to determine those who are qualified to take samples.

4.6.1   Understanding the chemistry.

The reasons a particular sampling and testing method was chosen need to be explained and understood. This explanation will frequently start with the more general concerns during sampling and then address the changes needed in the general plans when a variety of other conditions are encountered. This understanding is important to good sampling and the student needs to demonstrate understanding of the chemistry.

4.6.2   Sampling tools.

For most people the tools used in sampling are new and strange. The reasons for their design and their method of use need to be explained. Use of the tools should be a part of the instruction. Demonstration of the ability to use the tools should be a requirement to pass the training.

4.7   Sample Management

4.7.1   Labeling and records.

All samples should be labeled so that they are traceable to their bulk source, to the sampler, and to the date and time they were taken. Records must be accurately kept and must be identifiable with each individual sample.

4.7.2 Incoming samples.

Incoming samples should be stored so as to preserve their integrity before analysis. It is never safe to assume a sample received by the laboratory is homogenous. The sample should be mixed before analysis, if possible.

4.8 Retained Samples

4.8.1 Reasons for retaining samples.

- To establish a data base for a new or modified test.

- To establish a typical property not normally measured.

- To resolve differences between vendor and customer. Whenever possible, material from the actual containers involved should be tested and a retained sample used only as a backup or second choice sample.

- For periodic review of shelf-life data, product or process drift, etc.

- For reliability studies when changes have occurred in a process that cannot be related to any known factor.

- For compliance with the law.

4.8.2 Quantity of retained samples.

An amount at least twice as large as the amount necessary to run all the routine lot acceptance requirements is desirable.

4.8.3 Retention time.

The minimum retention time should be equal to the shelf life of the lot or one year after the final sale of the lot. Special environments may be required for long-term storage.

4.8.4 Container.

The container should be appropriate for the retention time and product characteristics under study. It is desirable for the container to represent the normal product package as closely as possible and to have no adverse effect on the material.

4.8.5 Security and environment.

A suitable system should be used to control access to retained samples to protect their validity and to preserve the original quality.

4.8.6   Hazardous products.

Retention of hazardous materials (e.g., flammable, volatile, shock-sensitive, toxic, or corrosive) presents a considerable risk, and this may influence retention plans.

4.8.7   Reactive products.

Many materials are reactive and will exhibit significant change in properties with time. For certain cases, such as elastomers, an "as is" sample should be kept for a short period and a "cured" sample kept for the usable life. The object of the latter is to validate that a good final product could be produced when the product was originally manufactured.

4.8.8   Regulatory requirements.

Any regulatory requirement which specifies retained samples shall be followed and be considered minimal good practice for any material manufactured under these requirements.

5.   Analysis and Testing

Scope:

- Any laboratory producing data for process control or final lot acceptance should meet the following requirements as a minimum for operation. The operation of the laboratory should be recognized as a process, and controlled and managed as such.

- Good quality practices should be defined for the documentation of test methods and general preferences established for international and/or national methods over nonstandard methods.

- Good quality practice for standards, controls, and traceability should be maintained. Calibration standards are often used to set test equipment to read correctly before being used in actual testing. Controls are used to monitor the performance of test methods.

- Certification of material can be expressed in several forms. The most common are Certificate of Analysis (Test) or Certificate of Compliance. The titles "Quality Control Report" or "Quality Assurance Report" are sometimes used to cover the same information. The documents are used to provide the customer with information regarding a given shipment.

  To be of value, a certificate must accompany the shipment or be sent under separate cover, arriving no later than two days after receipt of the product shipment. This timing should be negotiated with the customer.

5.1 Good Laboratory Practices

5.1.1 Personnel.

- Records should be retained covering:
  - Organization chart.
  - Job descriptions or other definitive declaration of individual authority and responsibility for all personnel.
  - Internal training of all related individuals.
- Types of individuals required to operate a laboratory:
  - *Technician.* Individual who carries out analysis and has demonstrated an ability to read, understand, and perform the prescribed procedures with documented training.
  - *Technical resource.* Individual with expertise and advanced training related to the prescribed procedure.
  - *Laboratory supervisor.* Individual responsible for coordinating daily activities, training, documentation, and the general running of the laboratory within company guidelines.

5.1.2 Documentation of operations.

- Records shall be retained covering:
  - Equipment maintenance.
  - Calibration and standardization including appropriate statistical control charts.
  - Corrective action activities related to out of control conditions.
  - Laboratory data with traceability to the technician who produced the results.
  - Test methods and laboratory procedures both current and historical with approval for change, revision dates, etc.
  - Sample schedules.
  - Laboratory responsibilities, authorities, and relationships.

5.1.3 Interlaboratory activities.

When two or more laboratories are testing or measuring the same characteristic on a given material, it is recommended that regular, controlled interlaboratory comparisons be conducted. The laboratories concerned may involve the supplier, customer, or regulatory agencies.

5.2 Test Methods Documentation

Analytical and test methods used for purposes of process control or to determine conformity to material specification requirements are central to the quality system

21

of a process industry. Each organization must establish procedures for the preparation, approval, and preservation of test method documentation. Exact procedure used should be well documented, including sampling, sample preparation, and conditioning, testing, recording, and preparation for next test (cleanup).

5.2.1 Compendia methods.

Compendia of test methods have been developed and compiled by several standards organizations such as the International Standards Organization (ISO), American Society for Testing and Materials (ASTM), Association of Official Analytical Chemists (AOAC), United States Pharmacopeial Convention, and others. Review processes and interlaboratory comparisons conducted by these organizations help to ensure the accuracy and reproducibility of the methods. Use of these methods, when available, is therefore recommended. During each test, environment, chemistry, and test equipment may differ and the standard method should be validated before use.

5.2.2 Nonstandard methods.

Deviation from the procedures specified by a compendial method are occasionally required due to limitations of available equipment or the properties of particular samples. When such nonstandard testing is to be conducted on a regular basis, or when legal or regulatory requirements are at issue, modified compendia methods should be documented as new methods by formal report.

5.2.4 Documentation of new methods.

Test methods other than established compendial methods must be documented by formal (usually in-company) report. Essential elements of the report for a test method include:

- Date of approval or last revision.
- Description of principle and validation of method.
- Range of applicability of test results using this method.
- All apparatus required for the test.
- Reagents required by chemical name and grade, Chemical Abstract Service (CAS) registry number, or by supplier and catalog number.
- Instructions for preparation of standard calibrants, controls, solutions, etc.
- Safety consideration and hazards of materials and equipment used in the test.
- Method of sample collection (see Section 4), protection, preparation, and conditioning.

- Amount of sample to be used, together with a stated tolerance on the amount (e.g., 5.0 +/− 0.1 g).

- Procedure for calibration of instruments.

- Procedure for controlling the test including sample control charts.

- Step by step instructions for carrying out the test procedure and calculations, including replication requirements, if any.

- Approximate time required for analysis.

- Test interferences if known.

- Precision statement on test repeatability and reproducibility, preferably in terms of standard deviations (as defined in ASTM D-3040).

5.3 Standards and Reference Control Materials Traceability

5.3.1 Traceability.

- Each testing organization (normally a laboratory) should designate appropriate primary standards (calibrants) for particular materials, weights, etc., required for its work. Laboratory supervision is responsible for taking precautions against their loss, contamination, or change.

- It is desirable that standard materials be traceable to recognized (e.g., National Bureau of Standards) sources. The essential requirement of good practice is consistency within and between the laboratories. If no recognized source is available, a mutually agreed secondary source may be used.

- When standards are used within a laboratory (for example, reagent solutions in ASTM E200), labeling for the standard should include at least:
  - Identity of the material
  - Reference to the laboratory notebook recording details of its preparation
  - Expiration date beyond which the material should be discarded

5.3.2 Standards versus reference controls.

- In some cases, primary reference standards have been developed by governmental agencies such as the National Bureau of Standards. It is good quality practice to use reference standards traceable to third party organizations or agencies when available. By definition, a reference standard has a unique given value, generally determined by an absolute method.

- More often, agreed upon internally generated reference materials or controls are used. These are often reagent grade or carefully selected large sample from a well-characterized and mutually agreed upon lot or batch.

These controls are used in many instances between customer and supplier. In general, they will have an agreed value and a confidence range. Reference materials (controls) are not calibrants.

- If a hierarchy of standards can be developed, that is, the measured values of the standard can be transferred mathematically from a primary material to a working standard or control, then this practice is acceptable to preserve what is usually a costly material.

- To measure drift with time and environment, reference materials (controls) are used to ensure the test method is still on target, by retesting at specified intervals. Control charts should be used. The test method should provide guidelines to be followed.

5.4 Certification

5.4.1 Certificate of analysis (test).

A certificate of analysis indicates that testing was done on a lot or shipment to a customer. The actual contents of the certificate should be negotiated between the supplier and customer. The following elements are suggested as part of the shipment identification:

- Supplier order/reference number
- Customer order number
- Test results
- Date of shipment
- Product identity
- Product quantity
- Product origin
- Name and title of the certifying individual

For the test results:

The actual test results for the specified lot should be given whenever possible. Supplying identification of the test method and the specification range for the tests is optional.

For various reasons, the test may be on other than the lot specified. It may be on a statistical or physical composite of lots, may be a periodic sample, etc. When this is the case, the test result may be reported on the certificate, but the source of the test value should be identified.

5.4.2 Certificate of compliance.

A certificate of compliance is an assertion that the product complies with a given specification. The certificate relates to the contents of a given shipment (see Section 5.4.1), but provides no test results.

5.5 Independent Laboratories

5.5.1 Purpose.

The purpose of this section is to present good practice guidelines for using independent laboratories.

Definition:

An independent laboratory in any testing or analytical facility which is not under the direct management control of the company using the services of such independent laboratory. Independent laboratories are used when particular tests or methods cannot be performed conveniently, technically, economically, or in a timely manner by the company needing the data or test results.

5.5.2 Accuracy and precision responsibility.

It is the responsibility of the requesting company to ensure that an independent laboratory has the capability, opportunity, and understanding necessary to meet the specific accuracy and precision requirements of the requested testing.

Prior to final commitment and at periodic intervals, the receiving company should fully verify that the results of the independent laboratory meet established accuracy and repeatability criteria.

- The company should provide appropriate reference samples for internal use at the independent laboratory. If a reference sample is subject to change with time or for other reasons, the receiving company should detail the necessary storage conditions and/or provide timely replacement reference samples. The receiving company should state the required repeatability on reference samples.
- The independent laboratory should periodically report the results of its statistical control data on reference samples to the receiving company. The results should be reported at least after 50 tests have been run on the reference sample.
- Blind samples or formal interlaboratory comparisons are appropriate means for auditing the quality partnership.

5.5.3 Qualification audit.

Before reaching any final agreement, the company should audit the independent laboratory to verify that:

- Adequate physical facilities and equipment exist to accomplish the test procedures.

- Testing staff are trained and capable.
- Sample retention facilities are adequate.
- Adequate statistical control procedures for test methods are in place.
- Proprietary information safeguards are adequate.
- Management fully understands the intended requirements.

5.5.4 Agreement on services to be performed.

There should be a written agreement which fully describes the entire expectations of the receiving company and which documents the agreement of the independent laboratory to meet those expectations.

- The written agreement should include, but not necessarily be limited to, the following:
  —Specific test methods to be used (ASTM or other)
  —Sample preparation responsibility
  —Sample identification details
  —Frequency of testing
  —Frequency of testing measurement control samples
  —Use of control charts
  —Repeat test requirements
  —Treatment of outliers
  —Location at which test will be performed
  —Details of sample transmittal
  —Details of result transmittal
  —Sample retention requirements
  —Protection of proprietary information
  —Record keeping
- There should be a detailed, written procedure covering retest, resample, and recalibration procedures.

5.5.5 The independent laboratory should be provided with the name and phone number (with backups) of a qualified contact in the receiving company who can handle technical or procedural questions.

6. Process Control

Scope:
- Good practice is described for process instructions and process documentation.
- The quality of the equipment used in a chemical process has a major impact on the safety of the process and the quality of the product. The equipment must be capable of carrying out the process as planned, safely.

- Technological changes (including procedural, test method, formulation, and specification changes) may be made to a process for reasons of improved product quality, improved safety, or improved process efficiency. Product characteristics may be altered as a result of technological changes. Technological changes should be carefully controlled and documented to assure that the product remains fit for the intended customer use.

- Good quality practice is documented concerning the use of statistical quality control in the CPI. These processes often make extensive use of dynamic (closed loop feed forwarded/feedback) control systems which are separate from the statistical quality control system. Statistical methods can also be built into the dynamic control systems.

- Quality is managed and assured in custom or research product manufacturing processes by controlling the tools and procedures of the process, together with a reliance on training of people.

- Process capability is defined and guidelines are provided for use of the concept.

6.1 Process Documentation

This section may be referred to as Standard Operating Procedures.

6.1.1 Written instructions must be used to describe in full the planned operation of any process. Quantitative terms should be used to describe all process parameters. Qualitative terms like "hot-cold" and "acidic-basic" are not sufficient to control a chemical process. Qualitative terms do not facilitate the application of statistics for process development, control, and improvement — essential elements of good quality written instructions.

Pre-startup decision points:

- Safety checklist and procedures either by explicit words or references to operating standards.
- List of raw materials to be used, including qualities, quantities, and inspections.
- Equipment inspection procedures.

Process operation:

- Step-by-step detailed description of the process:
  - Sequence of additions and process operations.
  - Points in the process where it may be placed on hold without affecting the quality of the product.
  - Parameters of the process in quantitative terms with target values and limits.

- Sampling schedule and procedures.
- In-process testing requirements and the actions which are to be taken based on the results.
- Container inspection and preparation procedures.
- Process records (log sheets, notebooks) for:
  -- Raw materials — quantity used and lot number
  -- Sequence and duration (actuals) of additions
  -- Parameters (actual) — temperature, pressure, stirring rates, flow rate, and duration
  -- Samples taken — when, where, and how
  -- Test results and actions taken
  -- All other process actions and observations not specifically listed

- Abnormal operation plan

  Instructions for proceeding in case of unplanned delays, emergency shutdowns, and other upsets which can and cannot be foreseen.

6.1.2 Responsibility.

The production department is responsible for the preparation of process instructions. The process instructions should be written by a team made up of production, product development, product engineering, safety, analytical, and quality personnel. Each should contribute his or her respective expertise so that a complete procedure will be provided to the production personnel.

6.2 Maintenance

6.2.1 Preventive maintenance programs are needed to maintain the quality and capability of the equipment. A preventive maintenance program includes:

- Scheduled shutdowns for equipment inspections to detect weaknesses before they cause an unplanned process interruption, especially for those areas in which inspections cannot be performed at other times in the process.

- Detailed procedures for shutdowns and checklists for inspections to ensure vital steps are performed and vital points are checked every time.

6.2.2 Repairs need to be made as soon as possible, preferably without interrupting the process operation or during planned shutdowns. Preventive maintenance promotes process and product uniformity/consistency.

- Parts need to be available to make repairs in a timely manner to minimize the duration of scheduled, and especially unscheduled, shutdowns.

- Skilled personnel are needed to perform the inspections and repairs.

6.3 Technology Change Control

6.3.1 Responsibilities.

- Since the manager at the site has ultimate responsibility for product quality, he or she has the responsibility for control of current manufacturing practice and for technological changes to current practice.
- Other company functions, such as product management, business management, quality assurance, engineering, development, and regulatory specialists should have input to technology change proposals.

6.3.2 Technology changes.

The following constitute technology changes which should be controlled:

- Formulation change
  - — Ingredients
  - — Amounts
  - — Sequence
- Process change
  - — Equipment
  - — Procedure
  - — Conditions
- Testing change
  - — Test limits change
  - — Test conditions change
  - — Test addition
  - — Test deletion
  - — Test frequency change
- Specification change
- Shelf life change
- Package component change
- Label revisions
- New raw material qualification
- Site qualifications
  - — Addition
  - — Deletion
- Sampling plan change

6.3.3 Customer notification.

- In general, customers should be notified of technology change because of the potential for unforeseen product quality or performance effects. The decision for customer notification is to be made jointly by marketing and manufacturing functions.

6.3.4 Documentation.

Agreed technology changes should be documented to provide traceability.

6.4 Use of Control Charts

6.4.1 Overview.

- Shewhart Control Charts (X, X-bar, R, s, MR) are often the best choice for process control if the required sampling is possible. They provide a widely understood framework of procedures to validate statistical process control.

- The Cumulative Sum (CUSUM) chart is frequently used for applications in the CPI where small shifts need to be detected as promptly as possible, the data are obtained one at a time, and computational aids are readily available.

- Attributes charts may be used in cases where variables data are not available. In chemical processes this situation frequently occurs in packaging and shipping areas.

- Standard variables control charts assume that the population is normally distributed and that samples are independent of each other. Many chemical parameters are in fact nonnormal. Many samples are autocorrelated with prior samples. Special treatment is often required.

- Control charts should be used to supplement and complement dynamic process control, not replace it.

- Control charts should also be applied to process and test equipment.

6.4.2 Policy.

To assure its fitness for use, the product must be made from a consistent supply of raw material which meets acceptance specifications. The product must be produced by a process which is consistent over time and passed by a statistically valid release testing procedure. A consistent manufacturing process is required to assure that product characteristics which cannot be tested are consistent from lot to lot. Statistical process control, using the discipline of control charts, is the recommended technique to assure a consistent process.

6.4.3 Control chart principles.

- Control charts represent a commitment to manage the process on "aim" and in a state of statistical control.

- The user has a commitment to investigate for special causes when nonrandom behavior is observed.

- The use of predetermined, statistically based limits determines the appropriate interval for control actions.
- Where appropriate for the process, averaging of a larger subgroup size may be used to increase a chart's sensitivity.

6.4.4 Parameters for control charts.

- *Aim (or target).* The aim is the goal or aiming point value for the process. It is often called centerline and is usually determined from either the midpoint of the specification range or the long-term mean for the process.
- *Standard deviation.* The total standard deviation of the subgroup size used for the control chart, not the standard deviation of units of product or the long-term variability. Each control chart procedure includes a method of obtaining this value. Some control charts use the range (times a constant) within a sample to develop an estimate for sigma.
- *Control limits.* Appropriate action must be initiated when the controlled statistic exceeds the action limit. Each control chart procedure includes a method of obtaining these limits. Multiple decision criteria may be employed. The use of several tests for abnormal patterns is helpful for successful application of the X, X-bar, R, and s charts.

6.4.5 *The Individuals (X) Control Chart* is a chart of individual values plotted in the form of a Shewhart X-bar chart for subgroup size n = 1. It is a widely used control chart in the chemical industry due to cost of testing, test turnaround time, and the time interval between independent samples. In many cases, the individuals chart is the practical choice because only single observations are available.

6.4.6 *The Moving Range (MR) Chart* shows the absolute difference between the current and "nth" previous values, and is often plotted in addition to the MA chart.

6.4.7 *The average (X-bar) chart* averages several observations together and is more sensitive to small shifts than the X chart. The control limits are narrower since the standard deviation of the sample subgroups are smaller than for individuals.

6.4.8 *The Range (R) Chart* is the absolute difference between the highest and lowest values within a subgroup. The R chart is nearly always plotted together with the X-bar unless the subgroup size is larger than 10 to 12.

6.4.9 *The Standard Deviation (s) Chart* is the standard deviation of the results within a subgroup. In the CPI it is appropriate to use the s chart instead of the R chart for subgroup sizes greater than 10 to 12.

31

6.4.10 *The CUSUM Chart.* The CUSUM is the sum of the deviations from target. The CUSUM can be implemented graphically or numerically. In the graphical form a V-mask is used instead of action limits. Whenever the CUSUM trace moves outside the V-mask, control action is indicated. In the numerical form, which is more widely used, the CUSUM is computed from a selected bias away from target value and a fixed action limit is used much like the Shewhart Chart. Additional tests for abnormal patterns are not required.

6.4.11 *The Moving Average (MA) Control Chart.* The moving average of the last n observations provides a convenient substitute for the X-bar in the Shewhart chart. However, it does require the user to specify a fixed number of samples, n, for averaging. Choice of the interval may be difficult to optimize. The control signal is a single point out of limits. Additional tests are not appropriate because of the high correlation introduced between successive averages which share several common values. The MA chart is sensitive to trends and is sometimes used in conjunction with x and/or MR charts.

6.4.12 *The Exponentially Weighted Moving (EWMA) Average Chart.* Each new result is averaged with the previous average value using an experimentally determined weighting factor, a. Usually only the averages are plotted and the range is omitted. The action signal is a single point out of limits. The advantage of the EWMA is that the average does not jump when an extreme value leaves the moving average. The weighting factor, a, can be determined by either an "effective time period" or by the method covered in Hunter's article (see Bibliography). Additional tests are not appropriate because of the high correlation between successive averages.

6.4.13 *The Multiple-Variable Control Chart.* Occasionally it is desirable to plot as the control chart variable a calculated parameter constructed from two or more measured variables (e.g., color or tensile tests). The calculated variable may be constructed on the basis of theoretical relationships, principal components, or Hotelling's T^2 (see Kotz, Johnson, and Read in Bibliography). The statistical properties of the combination should be validated before use.

Any of the control charts given previously can be used. Computational assistance is usually required. Attributes control charts (p, np, c, u) may be used, but are often less sensitive than variables charts. Fairly large sample sizes are necessary for good discrimination.

6.4.14 Sampling frequency for control charts.

- *Discrete process operation.* Good practice calls for sampling every batch if batch-to-batch control is desired. Within-batch sampling may provide additional information if the samples are independent and random. Batches may be skipped only after it has been shown that the process is stable over reasonably long periods.

- *Continuous operation.* Frequency of sampling is determined based on knowledge about the process and residence time. This period can be determined from process dynamics or from a statistical analysis of past data. Sampling periods may vary from 15 minutes to 24 hours or more. Intervals must be far enough apart so that samples are not auto-correlated.

6.4.15 Corrective action.

The occurrence of an action signal indicates that the pattern of the data is not consistent with the level and variability expected. Action should be taken to investigate and/or correct the process promptly. A list of appropriate actions should be available for immediate and uniform implementation. Emphasis should be placed on preventive action not merely compensating action.

6.4.16 Resampling/retesting.

Resampling or retesting not included in the design of the control chart should not be done. Resampling or retesting delays corrective action and changes the risk levels on which the chart control limits have been calculated.

6.4.17 Control chart implementation.

Standard Control Chart Applications

Control Variable	Threshold or Limit	Deviation or On-Aim
	— Type of Control —	
Process Average	X, X-bar, MA	CUSUM, EWMA
Variability	R, s, MR	CUSUM
Percent Defective	p	
Number Defective	np	
Number of Defects	c, u	CUSUM

See figure on page 34 for control chart selection.

- *Selection of variables.* The number of variables that can be maintained on control charts is limited by the facilities and the human attention span. It is generally not wise to put every measurement on a control chart. Priority should be given first to important product or process properties, and second to key control parameters. Charted variables should be controllable by the person(s) maintaining the chart.

- *Manual charting.* Control charts were developed prior to the wide availability of computers and all techniques can be implemented by hand

Guide for Control

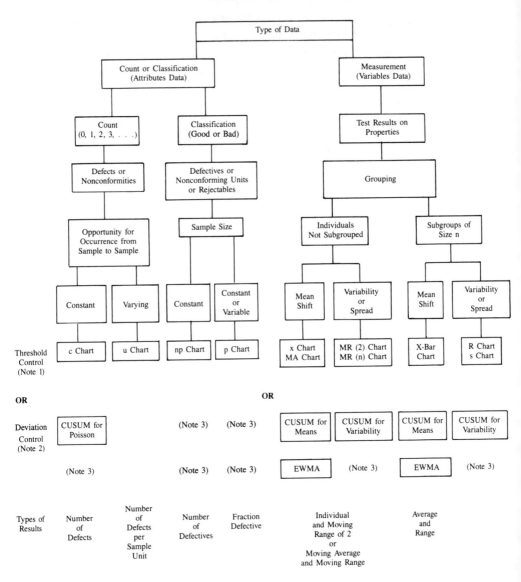

Note 1: *Threshold* Control indicates primary interest is in events near control limits. This type of control is most often used in the early stages of a quality improvement process.

Note 2: *Deviation* Control indicates primary interest is in detecting small shifts in the mean. This type of control is appropriate in advanced stages of a quality improvement process. Shewhart charts with runs rules may also be used.

Note 3: CUSUM for mean or EWMA may be applicable in these uses if data are transformed to approximate a normal distribution.

34

on paper. This is often the best choice for infrequent measurements, for pilot programs, or for short-term applications. Chart forms can be prepared in advance and operating personnel trained to plot the charts.

- *Computer software* is available for any control chart application. The use of computers can reduce the work, maintain a data base, allow the keeping of many charts and assist in diagnostic follow-up by use of graphical output on demand, adequate storage, error-correcting procedures, and checking of assumptions. Purchased programs should be validated before use.

6.4.18 Control chart maintenance.

Every control chart application should be periodically reviewed for performance and the estimates of the aim, standard deviation (or range), and the control limits modified as required. Charts with hourly sampling should be reviewed monthly and daily charts reviewed quarterly. The base period for the reference statistics should be at least three review periods if meaningful data are available.

6.5 Custom or Research Production Control

6.5.1 Introduction.

The chemical and process industries span a spectrum of production from continuous high volume products made routinely to specialty products produced in small quantities no more than once or twice a year. For specialty products, the application of good quality practices requires some specific consideration.

Often the production process is completed before sufficient samples can be tested and the results submitted. By the time the product is made again, the personnel and/or equipment may have changed and any process statistics would need to be redone. Thus, the concepts and applications of traditional process control and statistics are limited.

6.5.2 Process tools and conditions control.

Process tools and instrumentation require frequent standardization and statistical monitoring to demonstrate control (e.g., analytical instruments, scales, temperature controllers, flow meters, etc.). Their accuracy and precision must be determined. The goal is to reduce variation wherever possible.

Preventive maintenance programs are required.

6.5.3   Procedures control.

Process procedures are continuously subject to change. To minimize their impact on product quality, each must be rigorously evaluated (see Section 6.3). Changes for safety or changes in the analytical laboratory are process changes requiring similar evaluation.

6.5.4   Training.

Consistent, continual, current, and documented training is critical to control the variation contributed by personnel changes.

Personnel must display the skills and personal characteristics that support safety and quality.

6.5.5   Supplier approval.

Frequently, time does not permit the step-by-step supplier approval process needed for a new supplier. Long-term relationships with a few reliable suppliers are essential.

6.5.6   Process capability.

Process capability is frequently determined based on processes of a similar nature; e.g., all chemicals produced by a general type of synthesis.

6.5.7   Process development.

Process procedures most often have not been optimized. Experimental designs should be employed to improve the process continuously from run to run.

6.5.8   Sampling evaluation.

Agreement between customer and supplier is heavily dependent on the evaluation of proposed product sampling. To determine which tests are needed to measure acceptance of product, the communication between the customer and supplier must be well-established.

Sampling expertise based on product and process chemistry is very important.

6.5.9   Certificates of analysis.

The certificate of analysis is a major component of the communication process in custom manufacturing. Without the data to show long-term process control to targets, each batch must be individually characterized and the customer informed of the results.

6.6 Process Capability

6.6.1 General.

The process capability concept provides an important means for allocating resources and assessing the effectiveness of programs for quality control and improvement.

6.6.2 Definition.

Process capability is the calculated inherent variability of the product made by a process. It represents the best performance of the process over a period of stable operation.

Process capability is expressed as 6σ, where σ is the standard deviation of the process under a state of statistical control.

6.6.3 Process capability indices.

- Cp

  The Capability Index, Cp, is the ratio of the specification range divided by the process capability, 6σ:

  $$Cp = \frac{USL - LSL}{6\sigma}$$

  where USL is the upper specification limit and LSL is the lower specification limit. This σ is for the short-term component of variation.

  Values of Cp exceeding 1.33 indicate that the process is adequate to meet the specifications. Values of Cp between 1.33 and 1.00 indicate that the process, while adequate to meet the specifications, will require close control. Values of Cp below 1.00 indicate the process is not adequate to meet the specifications and that the process and/or specifications must be changed.

- Cpk

  The Capability Index Adjusted for the Process Average, Cpk, takes into account the possibility of the process average being off center. It is the difference between the process average and the specification limit nearest it divided by one half of the process capability, 3σ:

  $$Cpk = MIN \left[\frac{USL - PA}{3\sigma} , \frac{PA - LSL}{3\sigma} \right]$$

  where PA is the process average. This σ is for the short-term component of variation.

As for Cp, high values of Cpk are desired. Off-center operation will severely reduce the index and identify the need for an improved control of the process average.

6.6.4 Capability study.

- Evaluation period

Process capability should be computed from data collected from a period of steady operation in a state of statistical control (SOSC). If the requirement of statistical control is not precisely met, the process should be operated to a target value using the usual control techniques.

- Data evaluation

Control chart:

If the process is controlled using a conventional X-bar R chart, the process capability can be computed by the formula:

$$6\sigma \cong 6s = 6\frac{\overline{R}}{d_2}$$

Where s = the calculated standard deviation of the sample
$\overline{R}$ = the average range of at least 10 subgroups
d_2 = the constant for converting ranges into standard deviations for the particular subgroup size used.

Frequency distribution:

If no control chart is used on the property or if some other type of control chart is used, then a sample of at least 30 to 50 observations should be taken and the sample standard deviation, s, computed. The process capability is computed as:

$$6\sigma \cong 6s$$

where s = the calculated standard deviation of the sample.

6.6.5 Measurement variability.

Measurement variability is often much larger for properties important in the chemical industries (e.g., viscosity and purity) than it is for properties measured in the mechanical industries (e.g., dimensions and voltages). Measurement may be responsible for 50 percent or more of the observed variation.

Measurement variability may or may not be included in the estimate of the sigma used for computation of the process capability; however, it is important that the inclusion or exclusion of the measurement error be clearly intentional and specified in reporting the results.

If the measurement variability is large, care should be taken not to allow the use of sampling plans and sample averages to obscure true process variability. Consistent procedures should be used in computing the process capability for the same property and process at different times in order that the values be comparable.

6.6.6 Process performance.

Process performance should be distinguished from process capability. Process performance represents the actual distribution of product variability over a long period of time such as weeks or months, while process capability represents the product variability over a short period such as minutes or hours. Process performance variation will be wider than process capability since it will also include components due to time factors.

• Performance study when not in SOSC

It is sometimes desired to estimate the process performance capability over a long evaluation and/or when not in a state of statistical control (e.g., a customer requires that an index be reported even though control charts show out of statistical control results).

The Performance Index, Pp is:

$$Pp = \frac{USL - LSL}{6s}$$

where s is the calculated standard deviation for the total included population. The Performance Index Adjusted for the Process Average, Ppk, is:

$$Ppk = MIN[\ \frac{USL - PA}{3s}\ ,\ \frac{PA - LSL}{3s}\ ]$$

Pp/Ppk will always be less than or equal to Cp/Cpk, respectively.

6.6.7 Special problems.

• One-sided specifications

One-sided specifications present special problems. If the practice has been to run as close to the specification as possible, good quality practice requires that an operating target be set at least 3σ from the specification. If a capability index is computed, Cpk should be used.

However, there may be no nominal target value either because goal is as low as possible (e.g., impurities) or as high as possible (e.g., strength).

In this case, the process capability index concept should not be used to permit intentional contamination or degradation of the product.

- Nonnormal distributions

  The use of the 6σ factor in defining the process capability assumes that the distribution of the property follows the normal distribution. Routine plotting of histograms is a good practice to provide a graphical representation of process data. If it is known or suspected that the normal distribution does not describe a property, the investigator has the following four options:

  — Transform.

  Apply a transformation to the data to render the distribution of the transformed property effectively normal.

  — Known distribution.

  Use the multipliers which correspond to the 0.13 percent and the 99.87 percent points for the known nonnormal distribution.

  — Use as is as a relative measure and so note in report.

  — Omit the computation because the values so obtained would be misleading.

7. New Products/Processes

Scope:

Guidelines for good quality practices should be provided in the area of new products and new processes.

7.1 Sampling of Products Under Development

7.1.1 The development of a new product will typically involve sampling to potential customers and their testing of candidates prepared only in small pilot or preparative quantities. Full-scale process development and manufacture should begin only after such testing has created a market for the candidate product.

7.1.2 Important characteristics of the material which will be made from the scaled-up process are often unknown at the candidate stage. This situation makes close cooperation between supplier and potential customers especially vital, and they should communicate their respective product requirements (both as regards quality and quantity) and process capability on an ongoing basis during the development process.

40

7.2 Design Documentation/Process Transmittals

When the decision to commercialize a new product has been made, information necessary for detail design, startup, and later operation of production facilities should be recorded in a process transmittal. This documentation may be considered confidential. Among items to be included in a process transmittal are:

7.2.1 Product synthesis chemistry, thermodynamics, and kinetics.

7.2.2 Projected yields, time cycles, and process conditions at each step of the process.

7.2.3 Necessary information for process control, including adjustments to be used to meet key in-process and final product properties.

7.2.4 Physical and chemical properties of products, raw materials, and inter-mediates, including those related to toxicity, hazards, storage stability, and required materials of construction.

7.2.5 Tentative targets and limits for products and raw materials, together with test methods for these characteristics. It is recognized that specifications on a newly developed product may be based on limited data and require subsequent revision in the light of better determinations of customer requirements and of process capability.

8. Shipping

Scope:

- Guidelines should be provided for good quality practices in the area of label control for chemical products. This section does not specifically address the legal, safety, environmental, health, or regulatory aspects of labeling or label control. Container labeling control is aimed at in-suring that each container is correctly labeled as to the identity of the chemical it contains, the quality of that chemical, the quantity, and safety-hazard information required for the safe handling of the chemical.

- Good quality practice should be documented concerning the determina-tion and reporting of shelf life as it relates to chemicals.

- This section covers practices to ensure that product quality is maintained during loading, shipping, and unloading of chemical products from the producers to the customers.

41

8.1 Labeling Control

8.1.1 A container is any apparatus in which or through which a chemical is transported or stored including tank cars, hopper cars, trucks, bags, drums, bottles, boxes, pipelines, storage tanks, holding tanks, etc.

8.1.2 This practice applies both to individual containers which are the parent lot or any part of the parent lot. Accurate labeling is important at all points in the production, packaging, and handling of chemicals.

8.1.3 Packaged material.

Where possible, the exact number of containers needed to package the lot should be predetermined and segregated. If physical labels are to be applied to each container then the exact number of labels should be printed and the unique lot identifier included on each label. Otherwise, the label information, including the lot identifier, needs to be marked directly on each container.

8.1.4 Containers which cannot be physically marked, such as rail cars, trucks, etc., should be uniquely identified on records such as bills of lading and the certificate of analysis.

8.1.5 Pipelines of chemicals present labeling problems of their own. In those areas where the pipe cannot be relabeled in a timely or economical manner, labeling needs to indicate the category of chemicals carried by the pipe and the hazard warnings, i.e., flammable gases, flammable petroleum distillates, crude oil, process water cold — hot, etc. At the point of charge and discharge the specific chemical being transported by the pipe at that time should be clearly displayed. Labels at the charge and discharge points should include the purity and health and safety warnings for the chemical.

8.1.6 If a filling operation is interrupted and the product is still in conformance, a new lot number should be assigned before packaging is continued if serial marking is not being used, or the test results differ from earlier results.

8.1.7 Continuous production.

For continuously produced products the most effective system is to date-stamp the final containers because a lot only exists as a period of time. The exact number of labels is defined as a continuous supply for the duration of the packaging operation with each day's use equaling the number of containers filled.

8.1.8 Inventory records should include individual container serial numbers to allow segregation of portions of a lot.

8.1.9 There should be no shortage of labels which would leave a container without identification for any period of time. There should be no excess of labels which may be used on another lot or another chemical. All excess labels should be destroyed.

8.1.10 Consumer and industrial products.

Industrial chemicals are usually handled by trained personnel both at the manufacturer's plant and the customer's plant. Labeling of these products is usually more concise and technical. Consumer products will be handled and used most often by people with less of an understanding of the technical terminology and of chemicals. Consumer product labeling must be done with considerable care to ensure that the important aspects of a product are clear to the customer.

8.1.11 Minimum contents of a label.

- Name of chemical
- Health and safety warnings
- Net weight
- Lot number
- Company name and address

8.2 Shelf Life Control

8.2.1 Definition.

Shelf life is that period of time during which a chemical may be stored under specified conditions of temperature, relative humidity, and/or other physical conditions in the original sealed container and remain within the limits of the specifications for the chemical.

8.2.2 Determination of shelf life.

The determination of shelf life is an integral part of product development. For chemicals produced infrequently, adequate data is frequently not available to make a statistically sound determination of the shelf life. One should take into account the real world storage condition for shelf life determination for commercial products.

8.2.3   Extension.

- Since material may still be usable after the specified period of time has elapsed, the testing required to predict shelf life extension should be established by the manufacturer.

- The current owner of the chemical has the responsibility for the collection of any supporting data for the decision of suitability of the material for use, and for the actual extension.

8.2.4   Reporting shelf life.

The shelf life, where known to be finite, should be part of the labeling of the chemical. It can be reported via an expiration date or a date after which testing is required to reconfirm the quality of the chemical.

8.3   Product Shipment

8.3.1   Quality assurance of containers.

- The specific container to be used should be carefully selected with consideration given to corrosion, contamination potential, prior contents, in-transit conditions, loading and unloading facilities, product flow characteristics, and labeling needs.

- Procedures should be defined to assure the integrity of the container (to prevent possible loss or injury) and the cleanliness of the container. Inspection plans and responsibilities of both the container suppliers and the chemical product manufacturer should be defined and communicated.

- Dedicated containers may be desirable to reduce quality concerns.

8.3.2   Product loading for transport.

- New or reconditioned containers should be inspected prior to product loading to assure the integrity of the container and the absence of manufacturing residues, contaminants, or other foreign material.

- Handling procedures should assure that the potential for contamination is minimized. Sampling of dedicated containers may be required to determine if the heel will contaminate the incoming product. No product should be introduced into a container known or suspected to be dirty or which contains the heel of another product.

- Product packaged in drums, bags, or boxes to be shipped by truck or rail should have appropriate dunnage to prevent damage.

8.3.3   In-transit considerations.

- When possible, in-transit product transfers should be avoided. Contamination, damage, or loss are potential risks in these operations. If transfers (particularly from one container to another) are necessary, then the responsible carrier or terminal must be given clear and complete instructions for making the transfer.

- If product (or the container) is damaged or exposed to potential contamination, the owner of the product should be notified immediately. Under no circumstances should spilled or damaged product be repackaged and delivered to the customer without proper authority.

8.3.4 Product unloading.

- The receiver of the product should ensure that unloading lines and storage equipment will not jeopardize product integrity.

- If the filling operation is significantly interrupted on either a planned or unplanned basis, then samples may need to taken and tested to determine if the chemical is still in conformance with the specifications. The need for a resample and retest is based on the characteristics of the specific chemical and the specific filling process. Criteria should be established before the start of the filling operation.

- All interruptions need to be documented in the production records for later use in process control and problem resolution.

- A lot should be segregated before testing and not comingled with other production. If a rundown or load out tank is being placed into smaller containers, the larger vessel should be totally isolated from any streams or input which might change the quality of that lot.

- The assurance that the individual containers have not been contaminated during the filling procedure comes from process control procedures such as sampling the first and last container. The size and frequency of the filling operation determines the process control procedure needed.

9. Distributor Relations

Scope:

- Good quality practice should be documented concerning chemicals sold through or purchased from distributors.

9.1 Introduction

Good quality practices are equally applicable to the manufacturer, the customer, the transporter, and the distributor. The distinction between the quality programs needed by each lies in the product-handling functions (packaging, manufacturing, storage, filling, labeling, and bulk transfers) they perform and the responsibilities they have relevant to the quality of the chemical. The quality programs needed by these functions and the responsibilities are all well documented in "Quality Management and Quality

System Elements — Guidelines" and this publication. Each company must select the subgroup of these quality programs which will fulfill its responsibilities for quality.

The needs of distributors do warrant some discussion because their place in the supply chain has had little exposure or discussion.

9.2 General Distributor Quality Program Requirements

Appropriate services and expertise should be provided to suppliers and customers. A distributor should have as a minimum the following programs:

- Supplier Certification Program
- Shipping Control Program
- Storage Conditions Program
- Specifications Program
- Certificate of Lot Analysis Program
- Material Safety Data Sheet (MSDS) Program
- Transportation Requirements Program (for example, DOT)

Specific types of distributors may require more quality programs to support their business.

9.3 Manufacturer's Contract Sales Force

Certain distributors serve as the marketing function for the manufacturer. If the distributor handles the chemicals, they are in the sealed containers as received from the manufacturer. The manufacturer has responsibility for product quality until the chemical reaches the customer. The manufacturer must train the distributor in proper storage and handling of the products. The manufacturer should maintain a regular audit and training program to assure the distributor is capable of representing the manufacturer. The distributor needs the general distributor quality program listed in Section 9.2 to determine which manufacturers will provide the necessary quality support. The distributor must also implement those programs the manufacturer has determined are needed to maintain the quality of the chemical while in the distributor's possession.

9.4 Repackager and Contract Sales Force

Distributors who repackage product receive the chemical in large quantities and repackage with the manufacturer's approval and guidance into smaller packages. These distributors sell the product under the manufacturer's label. Here, the manufacturer again has responsibility for product quality. This distributor needs to add to his quality program a system for documentation of procedures for handling the products during transfer and subdivision, and a system for process control. Safety and disposal concerns also increase in this situation.

9.5 Manufacturer-Distributor

The manufacturer-distributor provides a combination of manufacturing and distribution. Products are received in bulk quantities, repackaged, and sold under the distributor's label. In addition, the products may be mixed further with other products to produce a new product. Also, the same product from different manufacturers may be stored and sold from a common tank. This distributor is responsible for product quality and needs a quality assurance program. This means a full quality management program of the type described by this manual.

9.6 Warehouser

This distributor serves as the warehouse and shipping arm of the manufacturer. The manufacturer arranges the sales. The quality responsibilities reside primarily with the manufacturer and are listed in Section 9.3.

10. Audit Practice

Scope:

Good quality practice should be documented concerning the internal or external audit or survey of quality-related issues within an organization.

10.1 Introduction

Quality audits are one of the major responsibilities of the quality function. They are a major and important tool in the establishment, measurement, maintenance, and improvement of quality. Quality audits determine the current status of decision making in an area and over time allow determination of progress and needs. Quality audits are equally applicable internally and externally. An audit is the objective quantitized measurement of an area's capability to do the job assignment. Auditing is applicable to all parts of the supply chain — manufacturer, transporter, distributor, and customer.

10.2 Types of Audits

There are as many types of audits as there are functions and products. Audits can be as specific as one function or production step or as general as a broad overview of the quality system for a whole company. The most commonly used generic types are:

10.2.1 Good Manufacturing Practice (GMP) audits get their name from the FDA audits which are regularly performed to determine compliance to the regulated manufacturing practices for pharmaceutical production. This type of audit is also used in the broad sense for determining whether a manufacturing process is able to control the product it produces.

10.2.2 Good Laboratory Practice (GLP) audits are also of FDA origin and assess the ability of a testing laboratory to provide accurate results for pharmaceutical products. Such audits are also applied in the broad sense to conduct an evaluation of a laboratory's ability to provide accurate results on any product.

10.2.3 Quality Systems audits determine the existence, structure, and health of the quality program in a company. This may include GLP, GMP, and evaluation of all support and manufacturing systems.

10.2.4 Product Specific Process audits start with the raw material selection for a process and follow every step of the process until the final product is shipped. They measure continuity, uniformity, and communication within the process.

10.3 Auditor Training

The primary prerequisite for an auditor is integrity and observation skills. Where possible these persons should have training in the ethics of auditing, e.g., an ASQC ETI course or equivalent will help to assure that the lead auditor has a complete program which does not overlook any important details.

10.4 Auditor's Position in Management Structure

The auditor's function needs to be defined to provide an unbiased performance of his duties. Each organization must make this decision and adjust as indicated by subsequent audits.

10.5 Conducting an Audit

The generic details of conducting audits are well documented. The specific program developed depends on the needs of the company.

Checklists are a valuable tool for conducting complete and repeatable audits. The sections of this manual are recommended as minimum points to be checked. Areas to be noted especially for CPI are:

- Cleanliness of the facility in all areas from production to laboratories and break rooms. Appropriate cleanliness is a must for the production of quality chemicals.
- A quality assurance function to interrelate separate departments and to assist nontechnical employees in comprehending their job's quality requirements.
- A testing laboratory operating in total quality control.
- A system for review, documentation, and control of changes to specifications for raw materials and products, procedures, test methods, and storage conditions.

- A quality improvement program.

- A sampling program which applies the most current knowledge of chemistry, statistics, and the production process.

- Adequate storage conditions.

- An evaluation of the commitment of the personnel to quality.

- A training program to assure the knowledge needed to perform a given job.

- A lot traceability system.

- Audit program existence, format, and schedule.

11. Customer and Supplier Relations

Scope:

- A description of good quality practices should be provided in the selection and approval of raw materials suppliers.

- Good quality practice should be documented concerning the sharing of product data with the customer. The data may take the form of a listing representing the material in the shipment as well as a chart representing some appropriate production interval.

- When quality problems occur, both the supplier and customer should focus on defining the problem, identifying the root cause, and planning action to prevent the recurrence. Information concerning both the supplier and customer's process that is needed for the resolution of the problem should be available.

- Any practice concerning a quality waiver should be equally applicable for intercompany shipments, intracompany shipments, inprocess intermediates, and product returns, and should be documented.

11.1 Supplier Approval

11.1.1 Customer-supplier relationship.

Because raw materials are major components of all processes, they can be a significant source of both process bias and assignable cause variation. It is highly desirable to consider the raw material supplier as a partner and develop a cooperative relationship. A confrontational relationship benefits neither the customer nor the supplier. While many supplier relationships involve sources outside the user's company, everything that follows can be and should be applied to intracompany suppliers as well.

11.1.2 Approval process.

- The identification of a prospective supplier includes consideration of the following factors:
  - Previous experiences and relationships with the supplier
  - Supplier's area of manufacturing expertise
  - Supplier's quality assurance program
- Once prospective suppliers have been identified, the proposed chemical specifications with their related test and sampling methods are provided. This is also an appropriate time to discuss, in detail, the supplier's quality system.
- Once the field has been narrowed by the previous two steps to the most qualified suppliers, it is recommended that discussions occur to share more detailed information concerning the raw material and the process needs of the customer.
- Representative samples can now be requested and evaluated for compliance to proposed targets and limits for further development of specifications, and for stability. They may also be use tested via bench scale reactions.
- Following laboratory evaluation of the samples, larger quantities of material may be utilized in customer's pilot plant trials, limited plant trials, or extended plant trials.
- At this point a quality audit of the supplier(s) is appropriate. Any prior audits of the supplier(s) are reviewed and updated where needed. An on-site audit may not be needed. See Section 10 for more details of audit practices.
- During this approval procedure, with the experience and knowledge gained, target and limits are refined.
- Based on the supplier(s) agreement to the specifications and completion of the above seven steps, the supplier(s) is approved.
- Note that the complete supply chain needs to be certified. If a distributor and/or transporter are a part of the chain then they too must go through the parts of the approval process appropriate to them.

11.1.3 Functions and responsibilities.

In most organizations the responsibility for the process of supplier selection rests with the purchasing function. It is essential that this be a coordinated team effort between purchasing, R&D, process development, process engineering, production, quality assurance, and the supplier. The responsibilities of these team members are listed as follows:

- Purchasing acts as coordinator of this team process, which brings together the potential supplier(s) and the user(s).

- Keeps communications current with the suppliers concerning the approval process, specifications, quantities, and schedules.
- Participates where agreed on in any audits of the suppliers.
- Maintains a list of approved suppliers.

• The supplier is responsible for providing information and data describing his product and process and for providing help to identify critical parameters that will ensure satisfactory performance.

• Product design functions (R&D, process development, and process engineering) evaluate candidate raw materials both at the laboratory and pilot plant levels and define acceptable ranges for specification limits.

• Production conducts final evaluations of raw materials and may perform in-plant production trials.

• Quality management coordinates the preparation of purchase specifications and leads the team which audits the supplier(s) facilities and quality systems. It also serves as the editor/author of the audit report and monitors the supplier's performance against the report and provides an overview of the approval process to assure all parts are kept in balance, i.e., a change in a test method is reviewed for impact on the specifications, etc. (see Section 6.3). Quality management assures all steps have been completed and thus forms a total process which serves as a basis for measurement of the relationship.

11.1.4 Single versus multiple sources.

While multiple sources may provide some assurance of a continuous supply of material, they can also result in greater variation in the final product. In processes where no variation (measured or unmeasured, known or unknown) is important, a single source is preferred. While the selection of a single source will provide decreased variation, the selection must be done carefully and should result in mutual commitment by the supplier and customer to the future of each other's company.

11.2 Data Sharing

11.2.1 Preparations for data sharing.

• A mutual discussion should be held by the parties involved before any data sharing is initiated. This discussion should address the following:

- What supplier data will be shared with the customer.
- What customer data will be shared with the supplier.
- How the data will be used including benefits to both parties.
- Availability of data desired by either party.
- Mechanism for data transmission including format and frequency.

- — What data is considered proprietary by either party.
- — Complete understanding of the nature and limitations of the data to avoid any misinterpretation.

- Security agreements may be needed if proprietary information is involved. In addition, the need for any special handling of such information should be clearly defined.

- Process control data

  - — Provides demonstration of effectiveness of Statistical Process Control.
  - — Most useful when presented in form of control charts or time sequenced raw data.
  - — Knowledge of amount of measurement variability in data is essential.

- Product property data

  - — Characterization of product as measured is provided.
  - — Knowledge of amount of measurement variability in data is essential.
  - — Accurate dating of data is required for product properties that change with time.
  - — Time relationship between supplier's process control, product property, shipments, and customer's preformance data may be needed to obtain full benefit from data sharing.

- General review of data sharing

  - — An agreed upon review of the data (quarterly, semi-annually, or annually) should take place for both the supplier's and the customer's mutual interest.
  - — Correlation between supplier's and customer's data may clarify specifications, problems, and the need for different data. This should be a working relationship.

11.3 Retrieval

11.3.1 Retrieval is the removal of material from the transportation and distribution network, and from customers, initiated by the manufacturer. The decision to retrieve material may be caused by manufacturer's concern over the safety or performance of the product, or in response to regulatory action. If a retrieval is a result of regulatory requirements, it is a "recall" and the specific requirements of the applicable regulations must be fulfilled.

11.3.2 Authority for initiating retrieval procedures should be clearly defined within the manufacturer's organization.

11.3.3 The success of a product retrieval depends on availability of complete records concerning the transportation and distribution of material which may be subject to retrieval. A manufacturer should maintain a system which will permit determination of the amount, date, and destination of all material of a given lot or series of lots, or all material made within a specified time frame. A distributor should keep records adequate to enable the customers of material from any lots distributed, together with the amounts and dates of shipment, to be identified.

11.4 Complaints and Product Return

11.4.1 A complaint report is a link between a customer's problem and those individuals in the supplier organization who can diagnose and correct the problem. It should cover not only obvious errors and failure to meet specifications but also performance problems, packaging, order problems, scheduling, and delivery problems, and inadequate specifications.

Complaints are measures of dissatisfaction. However, it should be borne in mind that a low rate of complaints does not mean product satisfaction. Study of complaints, while beneficial, is not a substitute for market analysis and field intelligence.

11.4.2 Classification.

Because complaints can and should cover many aspects of a product, a classification system based on the types suggested above should exist.

11.4.3 Elements.

The complaint report should contain the following elements:

- Identity.
  - Identity of the product/shipment involved
  - Product name
  - Manufacturing location
  - Shipping location
  - Shipping date
  - Lot number
  - Customer and supplier order number
  - Customer name
  - Customer location
  - Name of author of complaint report, date of receipt of customer complaint
  - Date of customer notification and any special requirements for response timing
  - Sample (if any) source and to whom sent
- Brief statement of product application.
- Complete statement of problem.

Relevant data and test results (with method identification) should be included. If product performance in an application is involved, lot numbers of satisfactorily performing material should be provided, if available.

11.4.4 Amount/value.

- Statement of the actual quantity of material involved.
- Value of claim or possible claim.

11.4.5 Sample of complaint material should be provided by the customer and should be fully identified and protected.

11.4.6 Response should contain the following elements:

- Direction to the individual who can perform or guide diagnosis of problem, determine cause, and recommend corrective action. This person is responsible for coordinating investigation and preparing response. Fast and complete response to a customer is imperative.
- Diagnosis of problem.
- Root cause of problem.
- Corrective action. (It would not be uncommon for the corrective action to be both long term and short term. If both are required, both should be reported.)
- Name of author of response.

11.4.7 Returned product.

Product may be returned for various reasons — complaint, shelf life, goodwill, etc. Disposition of returned product must be carefully considered and documented.

- Receiving department receives return and notifies appropriate individuals or department of receipt.
- Material should be inspected for packaging integrity.
- Materials review board (consisting of both production and nonproduction) should recommend disposition.
- Appropriate disposition should be taken: inventory, rework, discount sale, scrap, etc.

11.5 Quality Waiver

11.5.1 Shipment of product which does not meet specification is NEVER GOOD QUALITY PRACTICE. Any such shipment must be made under waiver.

It is necessary to clearly document the details of the shipment including the reason for the waiver and to inform all affected parties before the product is shipped.

11.5.2 If a product has been determined not to meet specifications for one or more specification properties, then the product can only be shipped if all the following criteria are met:

- For the properties which do not meet specifications, none are known to adversely affect the customer with respect to performance, reliability, or safety.

- The customer's end use is not adversely affected by the properties which do not meet specifications. This question should be answered by the customer whenever possible.

- Agreement to ship the product which does not meet specifications has been obtained from the customer or the marketing group responsible.

- The Material Review Board, if in place, has approved shipment as a proper disposition.

12. Bibliography

- ANSI/ASQC Q90-1987 *Quality Management and Quality Assurance Standards — Guidelines for Selection and Use.* Milwaukee: American Society for Quality Control.

- ANSI/ASQC Q91-1987 *Quality Systems — Model for Quality Assurance in Design/Development, Production, Installation, and Servicing.* Milwaukee: American Society for Quality Control.

- ANSI/ASQC Q92-1987 *Quality Systems — Model for Quality Assurance in Production and Installation.* Milwaukee: American Society for Quality Control.

- ANSI/ASQC Q93-1987 *Quality Systems — Model for Quality Assurance in Final Inspection and Test.* Milwaukee: American Society for Quality Control.

- ANSI/ASQC Q94-1987 *Quality Management and Quality System Elements — Guidelines.* Milwaukee: American Society for Quality Control.

- ANSI/ASQC Z1.4-1981 *Sampling Procedures and Tables for Inspection by Attributes.* Milwaukee: American Society for Quality Control.

- ANSI/ASQC Z1.9-1980 *Sampling Procedures and Tables for Variables for Percent Nonconforming.* Milwaukee: American Society for Quality Control.

- ASQC Vendor-Vendee Technical Committee. *How to Conduct a Supplier Survey.* Milwaukee: American Society for Quality Control, 1977.

- ASQC Vendor-Vendee Technical Committee. *How to Establish Effective Quality Control for Small Supplier,* 2nd ed. Milwaukee: American Society for Quality Control, 1985.

- ASQC Vendor-Vendee Technical Committee. *Procurement Quality Control,* 3rd ed. Milwaukee: ASQC Quality Press, 1985.

- Eagle, Alan R. "A Method for Handling Errors in Testing and Measuring." *Industrial Quality Control,* March 1954. pp. 10-15.

- Grant, E. L., and R. S. Leavenworth. *Statistical Quality Control,* 5th ed. New York: McGraw-Hill.

- Grubbs, Frank E., and Helen J. Coon. "On Setting Test Limits Relative to Specification Limits." *Industrial Quality Control,* (March 1954): pp. 15-20.

- Hunter, J. S. "The Exponentially Weighted Moving Average." *Journal of Quality Technology* 18 (1986): 203-209.

- Jackson, J. E. "Quality Control Methods for Several Related Variables." *Technometrics* 1 (1959): 359-377.

- Johnson, N. L., and F. C. Leone. "Cumulative Control Charts." *Industrial Quality Control,* (June, July, August 1962).

- Johnson, R. H., and R. T. Weber. *Buying Quality.* New York: Franklin Watts, 1985.

- Juran, J. M., ed. *Quality Control Handbook,* 3rd ed. New York: McGraw-Hill.

- Kirk-Othmer. "Sampling." Vol. 20 of *Encyclopedia of Chemical Technology,* 3rd ed. New York: John Wiley and Sons, 1982, pp. 525-548.

- Kotz, Samuel, Normal L. Johnson, and Campbell B. Read. "Hotelling's T^2." Vol. 3 of *Encyclopedia of Statistical Sciences.* New York: John Wiley and Sons, 1983, pp. 669-673.

- Lucas, J. M. "The Design and Use of V-Mask Control Scheme." *Journal of Quality Technology* 8 (1976): 1-8.

- Natrella, M. G. *Experimental Statistics.* National Bureau of Standards — Handbook 91, 1966.

- Nelson, L. S. "Interpreting Shewhart X-Bar Control Charts." *Journal of Quality Technology* 17 (1984): 114-116.

- Sahrmann, Herman. "1986/87 Annual Trend Forecast," ASQC Chemical and Process Industries Division Meeting, October 23, 1986. Also, *Quality Progress,* April 1987, p. 38.

- Schilling, E. G. *Acceptance Sampling in Quality Control.* New York: Marcel Dekker, 1982.

- Shewhart, W. A. *Economic Control of the Quality of Manufactured Product.* New York: D. Van Nostrand Company, 1931.

- Willborn, Walter. *Compendium of Audit Standards.* Milwaukee: American Society for Quality Control, 1983.

Quality Assurance for the Chemical and Process Industries — A Manual of Good Practices

Opinions and Comments

In the spirit of continual quality improvement we need your help and comments to keep this manual an active document.

What are the most valuable parts of this manual?

What are the least valuable parts?

Suggestions for improvements or additions (i.e., more detail, examples, condensation, deletion, new topics, etc.).

Would you or someone in your organization be willing to assist the committee in working on a revision or developing this into a quality standard through the ASQC Standards Committee? If so, please identify yourself and you will be contacted by the committee.

General comments about this manual.

Thank You!

American Society for Quality Control
Chemical and Process Industries Division
Chemical Interest Committee

Please mail to:

Production Editor, Quality Press
American Society for Quality Control
310 West Wisconsin Avenue
Milwaukee, Wisconsin 53203